YOUR KNOWLEDGE HAS VALUE

- We will publish your bachelor's and master's thesis, essays and papers

- Your own eBook and book - sold worldwide in all relevant shops

- Earn money with each sale

Upload your text at www.GRIN.com and publish for free

Bibliographic information published by the German National Library:

The German National Library lists this publication in the National Bibliography; detailed bibliographic data are available on the Internet at http://dnb.dnb.de .

Imprint:

Copyright © 2016 GRIN Verlag, Open Publishing GmbH
Print and binding: Books on Demand GmbH, Norderstedt Germany
ISBN: 9783668320079

This book at GRIN:

http://www.grin.com/en/e-book/341876/space-systems-solutions-for-disaster-management-in-nigeria-the-nigerian

Christian Anuge

Space Systems solutions for disaster management in Nigeria. The Nigerian experience

GRIN Publishing

SPACE SYSTEM SOLUTION FOR DISASTER MANAGEMENT IN AFRICA: THE NIGERIAN EXPERIENCE

Defence Space Agency Nigeria

Content

1. INTRODUCTION

Over the years, natural disasters have remained an enormous challenge to most countries in Africa. Incidents such as drought, famine, floods, oil spillage and wildfires often plague various countries across the continent. These disasters, particularly those related to meteorological, hydrological and climatic hazards are increasing across the continent. The situation is worsened by unplanned and unregulated land use, weak environmental controls, poor enforcement of building standards, urbanization, and other development-linked factors that increase the vulnerability of people, property, and infrastructure.

Moreover, a large portion of the African continent is not adequately covered by ground observation network. Consequently, it is difficult to manage natural disasters terrestrially. The use of space system is one pragmatic option that is yet to be fully explored in solving these problems. Nowadays, satellites have become invaluable tools for disaster management throughout the disaster management cycle.

This purpose of this article therefore is to carry out a systematic investigation and analysis of space solutions to disaster reduction and control in Africa. Space systems would improve measures relating to prevention, mitigation, preparedness, emergency response and recovery. The write up focused on Nigeria because the disaster pattern and management in the country bears some close similarities to the conditions in most other parts of Africa.

This article gives a vivid account of findings of the study of broad issues relating to disaster management in Africa using space systems and related technologies. This write up would start by reviewing the generic ways space systems can be used in managing disasters. Thereafter, it would examine the current status of space disaster reduction in Africa, with particular attention on Nigeria. Then it would also go ahead to look at some opportunities and challenges in African by unfolding some of the global efforts of space disaster management in the continent. The article would furthermore focus on Nigeria by carrying out some case studies on indigenous effort by NASRDA to use space to manage for checking desertification, floods and gully erosion.

2. SPACE SOLUTION TO DISASTER MANAGEMENT

2.1. This section would examine the concept of disaster management. Thereafter, it would give a brief description of how space systems could be used to proffer solution to disasters. Consequently, it would focus mainly on remote sensing technologies, satellite communications, Global Navigation Satellite System (GNSS) and meteorological satellites particularly in reference to early warning, vulnerability assessment, emergency preparedness, disaster mitigation, adaptive response, and disasters.

2.2 Disaster Management

A disaster can be defined as a serious disruption of the functioning of a community or a society causing widespread human, material, economic or environmental losses which exceed the ability of the affected community or society to cope using its own resources (C Westen , 2005). The consequences of disasters are enormous as it often leads to disruption of social and economic activities thereby endangering communities. In the past, disasters were considered as inevitable events which are beyond human control. In recent times however, disasters are been looked at as outcome of interactions between hazards and vulnerability. They are no longer perceived as sudden eruptions that are to be handled by emergency response and rescue services but as incidents that can be predicted, prevented or significantly reduced through effective disaster management.

Complementarily, management consists of decision-making activities undertaken by one or more individuals to direct and coordinate the activities of other people in order to achieve results, which could not be accomplished by any one person acting alone. It often requires a collective effort and resources of several individuals to accomplish a goal, which neither could achieve. Disaster Management could therefore be defined as the coordination and integration of all activities necessary to build, sustain and improve the capability to prepare for, protect against, respond to, recover from, or mitigate against threatened or actual natural or man-made disasters.

Disaster Management consists of a number of phases, each requiring a different range of activities ranging from preparedness, response, mitigation and recovery. The Disaster Management Cycle is illustrated in the diagram below.

Fig. 2.1: Disaster Management Cycle. Courtesy: S Ambrose et al

Although Africa may not be the most disaster prone continent, it is highly vulnerable to disasters because of physical, social, economic and environmental factors that negatively affect the capacity of the people to secure and protect their livelihoods. This largely due to factors such as poverty degraded environments, high prevalence of diseases and low access to social services, weak governance and armed conflict. The increasing population further worsens the situation as human infrastructures are getting more vulnerable to the natural hazards, which have always existed. The result is a dynamic equilibrium between these forces in which scientific and technological development plays a major role. Recurring occurrences of drought, floods, landslides and forest fires need to be studied using today's advanced technology to find effective preventive measures. Space technology offers an improvement to disaster mitigation process in Africa through better predictions; detection of disaster prone areas; location of protection measures and recovery plans.

2.3 Space Solutions to Disasters

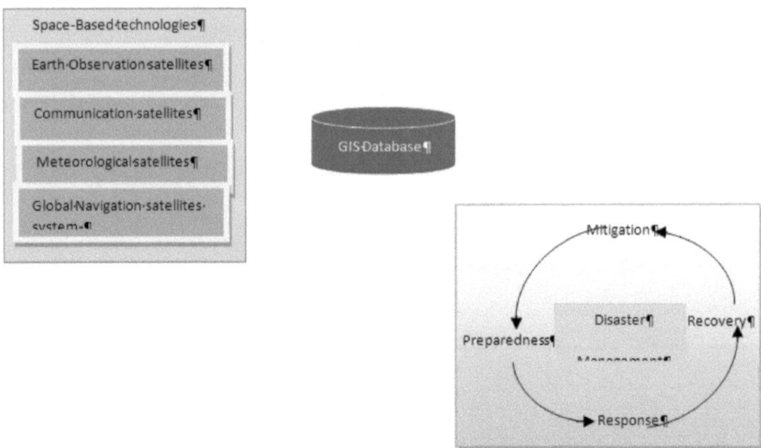

Fig 2.2 Space Solutions to Disaster Management.

2.4 Remote Sensing and Geographical Information System

Earth Observation System (EOS) used for Remote Sensing in conjunction with Geographic Information System (GIS) are among many tools available to disaster management practitioners. Satellite images enable disaster managers to have a broad overview of the environmental situation across a wide area. The scale of view from space ranges from entire continents to detail of a few meters. Thus, many types of disasters, such as floods, droughts, cyclones, volcanic eruptions, etc. will have certain precursors that satellite can detect. Remote sensing also allows monitoring the event during the time of occurrence. The vantage position of satellite makes it ideal for obtaining a big picture of the disaster thereby facilitating effective monitoring of the event

GIS is a system that captures, stores, analyzes, manages, and presents data that are linked to geographical locations. In the simplest terms, GIS is the merging of cartography, statistical analysis, and database technology. Thus, GIS is a necessary complement to Remote Sensing and it is useful at every phase of disaster management. In disaster prevention phase, GIS is used to manage the large volume of data needed for the hazard and risk assessment. In disaster preparedness phase, it is a tool for the planning of evacuation routes, for the design of centers for emergency operations, and for integration of satellite data with other relevant data in the design of disaster warning systems. In the disaster relief phase, GIS is extremely useful in combination with Global Positioning System in search and rescue operations in areas that have been devastated and where it is difficult to orientate. In the disaster rehabilitation phase GIS is used to organise the damage information and the post-disaster census information, and in the evaluation of sites for reconstruction. Hence, GIS is the useful tool in disaster management if it is used effectively and efficiently (Pearson et al., 1991).

2.5 Satellite Communications

The role of communication services in managing disaster emergencies cannot be over-emphasized. Effective communication enhances the success of disaster mitigation and relief operations which is often dependent on teamwork. Also, in the event of a disaster, satellites are the only wireless communications infrastructure that are not susceptible to damage , because the payload and main repeaters sending and receiving signals are far away on the Earth's orbits.

In addition, satellite communications networks provide wide coverage wireless connectivity in support of support emergency response activities. They are capable of providing a full range of communications services, including voice, video and broadband data. Communication Satellites operate with ground equipment ranging from very large fixed gateway antennas down to mobile terminals the size of a cellular phone.

Another significant merit of satellite communication system is its ability to provide connectivity over vast areas that cannot be linked by terrestrial networks. Moreover, satellite communications are able to connect difficult terrains such as marshy grounds, rocky mountainous and geo-graphic areas where it would be difficult to use alternative technologies.

Other advantages offered by satellite communication systems are:

- Wide area coverage that cuts across cuts across political or national boundaries.

- Facilitates mobile and wireless communication independent of location, demographics and terrain.
- Wide bandwidth available throughout coverage area.
- Communications cannot be obstructed by terrestrial infrastructure.
- Speedy deployment and recovery of ground equipment in case of emergency.
- Relatively low marginal cost per added site.

From the foregoing it is obvious that many of the unique communication needs for managing disasters in the African continent so closely fit the advantages offered by satellite communication systems.

2.6 Global Navigation Satellite System

In disaster management, GNSS is essential for positioning and navigation. For instance in Search and Rescue operations the location of distress vessels such as ship or aircraft is a crucial aspect in the event of a mishap. GNSS can be used to provide such critical positional information. GNSS can also be used for referencing of image for maps generation and updating of GIS for disaster management. Furthermore, mapping and field assessment can be done by GNSS. Other uses include coordination and monitoring vehicles involved in disaster operations.

GNSS could therefore be a useful means for Africa to be in tune with the current trends in ICT and accurate spatial information gathering. This space technology has an enormous potential to contribute to the management of environment, natural disasters, emergency response, improve the efficiency in surveying and mapping. It would also guarantee safer navigation to avoid maritime and aviation disaster.

Search and Rescue (SAR) initiatives like COSPAS-SARSAT make use of satellites to receive and transmit distress signals. GNSS core systems like the European GALILEO are also developing the Search and Rescue GALILEO Mission (SARGM) as a European contribution to the international cooperative efforts in SAR activities. SARGM is designed to fulfil the International Maritime Organization (IMO) and International Civil Aviation (ICAO) regulations and requirements in terms of detection of emergency beacons.

2.7 Meteorological Satellites

Meteorological Satellites (METEOSAT) are used to provide information for predicting areas of flood and weather conditions for disaster prevention. Some of the observations' systems within the Integrated Global Observing Strategy Partnership, like the World Weather Watch Global Observing System (WWW-GOS), the Global Atmospheric Watch (GAW), the World Hydrological Cycle Observing System (WHYCOS) and the Global Climate Observing System (GCOS), derive their information via the use of meteorological satellites. The use of these satellites helps in predicting areas of flood and also weather conditions for the aviation industry

3. CURRENT STATUS OF AFRICA IN SPACE DISASTER CONTROL

3.1.   Over the years, Africa has been receiving global attention in various aspects of disaster control. There are several notable international organizations taking advantage of space systems in managing disasters in the continent. Also, a few African countries are beginning to venture into space applications in disaster control. This chapter would examine the current status, challenges opportunities and future prospects of space technology in Africa, particularly as it relate to disaster management. Before attempting discuss these challenges and opportunities, it would focus on some notable international organizations dealing with disaster related issues in Africa, starting with the UN-SPIDER.

3.2 GLOBAL INITIATIVES OPERATING IN AFRICA

3.2.1 UN-SPIDER

The UN-SPIDER is the United Nations platform for Space-based Information for Disaster Management and Emergency Response (SPIDER). It was established by the UN resolution 61, section 20, of 2006, with a mission "To ensure that all countries, international and regional organizations have access to and develop the capacity to use all types of space-based information to support the full disaster management cycle" (*UNOOSA* 2006). It was set up to provide universal access to all types of space-based information and services relevant to disaster management.

Since its establishment, it has become a gateway to space information for disaster management support which serves as a bridge to connect the disaster management and space communities. It has also been a facilitator of capacity-building and institutional strengthening *(CIS, 2008).*

3.2.1.1 UN-SPIDER in Africa

UN-SPIDER maintain worldwide networks of National Focal Points (NFP) locations in certain countries. In Africa the NFPs are located in Burkina Faso, Burundi, Egypt, Ethiopia, Kenya, Malawi, Mauritania, Mauritius, Morocco and Tanzania. In addition, the UN-SPIDER maintains Regional Support offices in Algeria, Nigeria and South Africa. The NFPs are national institutions that work with UN-SPIDER staff to strengthen national disaster management planning and implement specific national activities that incorporate space-based solution in Disaster management *(UNOOSA 2009).*

Some of the objectives of the UN-SPIDER in the Africa are to:

a. Establish a network to ensure that all countries in the regions have access to space-based information needed in disaster management.

b. Build capability in the countries in the regions in using space-based solutions to effectively support the management of disaster events, and coordination.

c. Create of a forum to bridge the gap between the disaster management and space communities to enable optimal use of the existing resources of the space agencies

d. Involve the disaster-management communities in developing user-driven projects for the benefits of the user communities in the regions.

e. Set up of outreach activities, and developing knowledge management and transfer, through case studies, best practices to identify projects and implementation

3.2.2 International Charter on Space and Major Disaster

The International Charter on Space and Major Disasters (ICSMD) aims at providing a unified system of space data acquisition and delivery to those affected by natural or man-made disasters through authorized users. Each member agency commits resources to support the provisions of the Charter thus assisting to mitigate the effects of disasters on human life and property. *(Bhanumurty & Shrivastava,2010)*. ICSMD is a collaboration of world major space agencies that have made their satellites available for humanitarian purposes by acceding to the International Charter on 'Space and Major Disasters'. The member agencies are ESA, CNES, CSA, ISRO, NOAA , CONAE , USGS, JAXA , and DMCII .

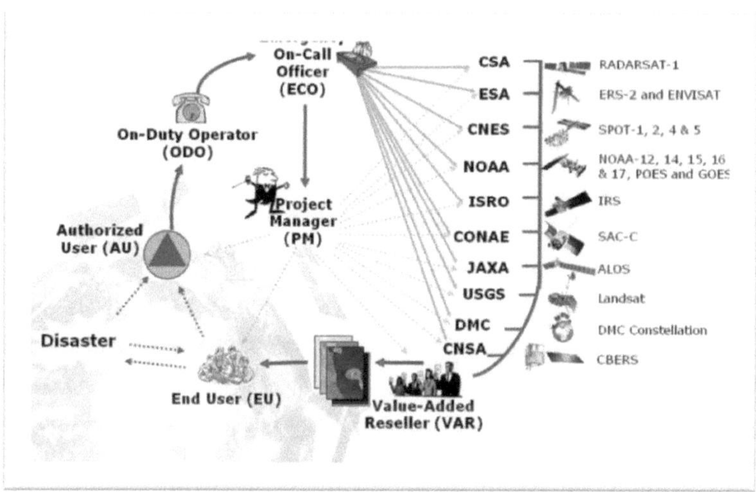

Fig 3.1: ICSMD Operational Loop (Setinel Asia, Bankok, 2010)

Although African countries are not among the Charter members, disasters in Africa have been amply covered as a result of close cooperation between the Charter Authorized Users (AU) and the UN agencies represented by the United Nations

Office for Outer Space Affairs (OOSA) in its capacity as a charter cooperating body. The AUs are the national civil protection authorities of the member states. In the near future, it is likely that African countries like Algeria and Nigeria, which are part of the Disaster Monitoring Constellation (DMC) with their own satellites (ALSat-1 and NigeriaSat-1), may become Charter members and thus may be able to activate the Charter directly. It is hoped that cooperative arrangements with the Charter Authorized Users would be maintained and that the African communities would continue to benefit from the space data and information for disaster relief channeled through the existing African space agencies.

3.2.3 Africa Earth Observation Network

Africa Earth Observation Network (AEON) is a research and teaching network of African and Africa-based scientists with its head office located in Cape Town. AEON is concerned with scientific understanding of how our earth works; the resilience of its interconnected systems; the value of its services; and documentation of its heritages, particularly in Africa

The primary goal of AEON is the cultivation of a high-level, internationally-connected, scientific research and learning environment promoting a modern interdisciplinary view of our Earth, and of Africa in particular.

3.2.4 African Association of Remote Sensing of the Environment

The African Association of Remote Sensing of the Environment (AARSE) was formed in August 1992 during an international conference organized by the United Nations in cooperation with the Government of the United States of America on "Satellite Remote Sensing for Resources Management, Environmental Assessment and Global Changes Studies: Needs and Applications of Developing Countries" held in Boulder, Colorado, USA. *(AARSE, 2010)*

The objectives of AARSE among others are:

a. To assist its members as well as national, regional and international user community through timely dissemination of scientific, technical, policy and program information in all aspects of space science and technology.

b. To create for addressing issues of common interest through the conduct of conferences, seminars and workshops.

c. To promote a greater cooperation and coordination of efforts among African countries, institutions and industries in the development of space technology and its application to natural resources and environmental issues.

d. To promote greater appreciation of the benefit of the technology, especially, remote sensing and Geographic Information System (GIS) in the pursuance of an African priority program for Economic Recovery and sustainable development.

e. To exchange views and ideas on technology, systems, policy and services of remotely sensed data and GIS which are applicable to the betterment of Africa.

f. To improve teaching and training in remote sensing and GIS and to collect, evaluate and disseminate results and failures in remote sensing activities from all over the world.

g. To conduct other remotely sensed and GIS activities consistent with the goals of AARSE.

Over the years, AARSE has worked closely with many international institutions and organizations such as the UNOOSA, UNESCO, UN-FAO, IEEE Geoscience and Remote Sensing Society (IEEE/GRSS), the Canadian Space Agency (CSA), European Space Agency (ESA) and NASA. In 2004, AARSE became an Organizational Member of GEO and has contributed towards the implementation of GEOSS. In 2005 and 2007, AARSE held two jointly organized (AARSE-ISPRS-IEEE-OGC-University of Johannesburg) GEOSS Workshops, in South Africa and Burkina Faso. Its active participation in the International Council for Science (ICSU) - Regional Office for Africa, UNESCO-GOOSE Programme, GSDI, AFREF, UNECA CODI-Geo, UNEDRA and other initiatives in the African Continent and the open support it enjoys from various remote sensing and GIS institutions and governments in Africa is a reflection of its success and commitments to the betterment of Africa (Ibid).

3.2.5 Famine Early Warning Systems Network

The Famine Early Warning Systems (FEWS) is a system for collecting, organizing and analyzing information relevant to food access and availability, in a number of African countries, in order to promote informed decision making by policymakers. FEWS is funded by USAID and it monitors food security via satellites. It uses vegetation indices, calculated from sensors including Advance Very High Resolution Radiometer(AVHRR), Moderate Resolution Imaging Spectroradiometer (MODIS) and those aboard SPOT, to monitor vegetation vigour and density and spot problems as they develop (Lewis S, 2009). It estimates rainfall using Meteosat infrared data, combined with rain gauge reports and microwave satellite observations, so as to model hydrological systems and how weather patterns might affect agriculture. The FEWS project has been an important user of remotely sensed information provided by NOAA and NASA. Apart from FEWS NET, another initiative using satellite data for drought and famine prediction and monitoring is the Global Monitoring for Food Security GMFS, funded by the European Space Agency (ESA).

3.2.6 Global Monitoring for Food Security

Global Monitoring for Food Security (GMFS) provides early warning, agricultural mapping and crop yield assessment services in support of food security monitoring activities in Africa It brings data and information providers together, in order to assist stakeholders, nations and international organizations to better implement their policies towards sustainable development *(GMFS 2010)*. GMFS collaborates with other key actors in the sector at the international, regional, and national levels. In Africa, GMFS activities at national level focus mainly on Ethiopia, Sudan, Senegal, Zimbabwe, Mozambique and Malawi.

3.2.7 Advanced Real Time Environmental Monitoring and Information System

The Advanced Real Time Environmental Monitoring and Information System (ARTEMIS) have been monitoring seasonal growing conditions and vegetation development over Africa since 1988. Since then, the Food and Agriculture Organization of the UN (UN-FAO has been using data from low-resolution satellites to monitor rainfall and vegetation conditions over large areas. These data are provided operationally in near real-time through the FAO-ARTEMIS. Primary users ARTEMIS data are the FAO's Global Information and Early Warning System (GIEWS), the FAO Emergency Centre for Locust Operations (ECLO) and the FAO Agrometeorology Group. *(FAO, 2005)*

Originally, the focus of ARTEMIS was on Africa and the system provided Cold Cloud Duration and Estimated Rainfall Images derived from the METEOSAT satellite and NDVI images from the NOAA-AVHRR sensors. Over the past few years ARTEMIS has increased its geographic coverage, as well as its range of products. A variety of image software applications for display and analysis have been developed in tandem with the ARTEMIS data products. These are now widely used in both developed and developing countries and, in particular, at several national and regional organizations involved in early warning for food security. (Michele Bernardi et al. 2000)

From the foregoing, there is a recognition that space is an essential tool for decision making and it is a useful tool for disaster management in Africa. Therefore, African national governments are expected support international space initiatives by setting up platforms and activities aimed at creating access to space-based information for disaster management. Effort could also be made to bridge the gap between disaster management institutions and existing African space centres so as to facilitate capacity-building and institutional strengthening in space disaster management. Also, establishing accessible linkages to existing facilities and archives could greatly contribute to continental sustainability

3.3 Challenges of Space Disaster Control in Africa

Africa has received prompt global and regional attention in space disaster prevention, management and reduction. Even though a few African countries are beginning to venture into space applications, the use of space-based systems in disaster prevention, reduction and management still remains a major challenge in Africa. Thus, natural disasters are still poorly managed leading to great economic loss to the region. The inability of many countries of the continent to effectively manage disasters on their own can be adduced to factors such as improper planning, insufficient government funding and dependence on archaic or non-functional systems of disaster management. This report would now elaborate more on 3 space related factors factors namely; lack of digital maps, non utilization of satellite data, and unplanned urban development.

3.3.1 Lack of Digital Maps

Many African countries still rely and use printed paper maps instead of digital maps for managing disasters. For instance, paper chart are still used for plotting ships' position, at NIMASA. In some African countries, the process of digitizing the bulk of hazard and regulatory maps is still inadequate. Even where available data is stored, they are scarcely shared between agencies or even between neighbouring states (K Adepoju 2010).

3.3.2 Non Utilization of Satellite Data

A few African countries such as Nigeria and Algeria have earth observation satellites, capable of generating remote sensing data. Unfortunately, such satellite data are not adequately adopted by emergency response agencies on the continent for planning, monitoring and damage assessment. Some agencies do not know or may not even have access to relevant global organizations and their functions in disaster management (K Adepoju, 2010). Therefore, the Africa space resource is highly underutilized for disaster management.

3.3.3 Unplanned Urban Development

Another major challenge to the ability to make progress in disaster management and crisis response in Africa is unplanned urban development. According to the special representative of the secretary general for Disaster Risk Reduction in Africa, Ms Margareta Wahlstrom " the rapid unplanned urbanisation in Africa creates high risk patterns and expose an increasingly large proportion of the population to floods, landslides, epidemics and other hazards." Wahlstrom also observed that most African governments have not critically assessed and improved vital infrastructures like hospitals and research centres, which play a key role in disaster management and crisis response in Africa. In addition, there is lack of appropriate mechanisms and coordination among disaster management agencies and organs. With this follows the difficulties of information flow and simple means of accessing information. Therefore, when a disaster strikes many African countries are caught unprepared and unable to respond. (Margareta Wahlstrom, 2010).

3.4 Prospects of Space Solutions in Africa Disaster Management

In spite of aforementioned challenges, some African countries have recorded some marginal improvement in the area of space application to disasters. In recent past, a few African countries have risen to the task of disaster management and reduction using space systems. The prospective areas of improvement vary from search and rescue services connected with aviation and marine transportation-related disasters to earth-observation data services such as the use of remote sensing and meteorological satellites data. In Nigeria for example, disaster management institutions such as National and States Emergency Management Agencies, the Nigerian Meteorological Agency and the Nigerian Maritime and Safety Organization have all embrace the use of space systems and are working in collaboration with NASRDA effectively.

3.4.1 African Partnership in the Disaster Monitoring Constellation

Nigeria and Algeria are key players in DMC constellation. The DMC satellites which include Alsat-1 and NigeriaSat-1, are already been utilized to monitor disasters in Africa and even globally. For instance, the DMC satellites were used in the acquisition of images during the Asian Tsunami disaster. Notably, NigeriaSat-1, acquired over 20 images which were downloaded to the central archives of the DMCII. These images were made available to RESPOND, an initiative created by the Global Monitoring for the Environment and Security (GMES), which is responding to emergencies using dedicated space-based applications. The DMCII has also provided satellite services for earthquakes and flood disasters in Iran and Philippines respectively. Furthermore, DMCII data have been used extensively for relief management and refugee settlements in the Dafur area of the Sudan. *(Momoh J and Akinyede, 2008)* The proposed NigeriaSat-2, which is an improvement on NigeriaSat-

1, with a higher resolution of 2.5-m panchromatic and 5-m and 32-m multi-spectral in four bands, will be more useful to the continent and other parts of the world in disaster management when it is launched.

The DMC will be complemented by the proposed African Resource and Environmental Management (ARM) constellation satellites. The ARM initiative was proposed by South Africa and supported by Nigeria and Algeria. The ARM project was conceived with the goal of building on indigenous knowledge to develop and transfer satellite technology by means of joint participation and knowledge sharing. It is part of the ideals of the New Partnership for Africa's Development (NEPAD) to provide a platform for scientific excellence in Africa in order to be globally competitive and contribute to the socio-economic development of the continent. The space segment of the ARM project will consist of identical satellites with panchromatic and multi-spectral payloads of 2.5-m resolution and 5-m resolution in six channels respectively. The satellites will afford Africa access to rapid, unrestricted and affordable access to satellite data for resource and disaster management.

3.4.2 COSPAS SARSAT Mission Control Centres in Africa

Nigeria, South Africa and Egypt have acquired the relevant facilities and established COSPAS SARSAT local user and mission control centres. The importance of COSPAS SARSAT programme for Search and Rescue especially in aviation and marine transport operations cannot be over-emphasized. Nigeria, South Africa and Egypt are in the right stead to be the hub for distributing alert distress data in West, South and North African sub-regions respectively.

3.4.3 Space Geodetic Techniques for Coastal Dynamics Monitoring

Space geodetic techniques are currently used to monitor the movement of the Earth's crust which usually give rise to earth tremors, volcanic eruption and coastal dynamics. Only a few related programmes and institutions exist in this field in Africa, probably due to high cost of acquisition, installation and maintenance of the relevant facilities. Also, there are few experts on geodetic technique in the continent.

In Nigeria, for example, there is a Centre for Space Geodesy and Geodynamics on Toro, Bauchi State. The Centre was established to facilitate the emplacement of the appropriate facilities and programmes for monitoring of crustal deformation and coastal dynamics in the West African sub-region. In addition, South Africa is participating in many of the existing schemes established around the world while Gabon and the Niger Republic are involved in the Doppler Orbitography and Radio-positioning Integrated by Satellite- Global Positioning System (DORIS-GPS) *[Ibid]*.

3.4.4 Remote Sensing Infrastructures in Africa

Remote sensing infrastructures, in most countries on the African continent, are still at a rudimentary stage. However, few countries such as Nigeria and Algeria have established space programs and launched imaging satellites with technical help from the United Kingdom. NigeriaSat-1 and AlgeriaSat are part of the Disaster monitoring constellation (DMC).The DM consists of a number of earth observation satellites constructed by Surrey Satellite Technology Limited (SSTL) and operated for the Algeria, Nigeria, Turkey, UK and China.

Other countries such as Kenya are planning on installation of infrastructure for reception of satellite imagery. The focus of most institutions and governmental agencies appears to be on applications. Understandably, this is a situation driven by necessity. As a result, little effort is given to theoretical studies and research in the remote sensing field, particularly in the areas that would benefit Africa the most. Even the existing human, technological and institutional capacity of the national space agencies in Africa are insufficient to permit full use of the latest remote sensing theories, techniques and applications. (Rochon G L et al 2005)

Some space related organizations involved in remote sensing in Africa are:

- African Regional Centre for Space Science and Technology Education in English (ARCSSTE-E), Nigeria.
- African Regional Centre for Space Science and Technology in French (CRASTE-LF), Morocco.

- Agence Spatiale Algérienne, Algeria.
- Centre Royal de Télédétection Spatiale Morocco.
- National Space Research and Development Agency Nigeria.
- SADC South Africa.
- National Remote Sensing Center of Tunisia.

Despite the existence of these of institutions, many remote sensing entities in Africa still lack access to current remote sensing products such as data, models algorithms and other value added services which address latest advancements in the field. The establishment of robust operational remote sensing or geospatial information systems and the supporting infrastructure are critical to securing timely access to reliable data and accurate information necessary to manage disasters. Therefore, African countries must consider synergy of effort in this direction by pooling resources at the local, regional, and continental levels. In addition, there is a need to develop and enhance the present infrastructure for remote sensing and GIS including, the organizational and computational data processing, particularly as relates to anticipated disaster mitigation needs in Africa.

3.4.5 Satellite Communication Infrastructures in Africa

Due to the advantages of satellite communications, the use of satellite capacity is growing steadily in Africa. This growth is increasingly market driven due to the demand for affordable broadcasting services, access to the internet, corporate data services and market competition in the industry. Hence, the proliferation of VSAT antennas is easily noticed on building structures around African cities such as Nairobi, Dar-es-Salaam, Kampala, Kigali and Lagos amongst others. On the commercial scale, satellite communications is therefore improving in Africa.

However, the improvement is yet to impact on disaster management needs across the continent as there is still shortage in the demand for capacity. For instance, C-band down-link capacity into Africa is still limited, with any spare capacity being used up by Internet Service Providers (ISPs) for internet connectivity. Up till now, the major sources of satellite capacity in Africa are non indigenous providers such as Intelsat, PanamSat, Intersputnik, Eutelsat and EuropeStar to name a few. These foreign firms are purely commercial and may not want to invest in less profit oriented venture such as disaster control. There is therefore the need for Africa countries to develop indigenous additional capacity over Africa for public utility needs such as disaster reduction.

3.4.5.1 Nigerian Communication Satellite

Few African countries like Nigeria are beginning to make indigenous effort in provision of satellite communication services. Nigeria launched a communication satellite (Nigcomsat-1) in May 2007. Nigcomsat-1 "footprints" cover Africa, parts of Middle East and Europe. The communication satellite is a hybrid geostationary satellite designed to operate in the C, Ku, Ka and L bands with a payload of 40 transponders.

It is exclusively designed to meet the current and future needs of Information Communication Technology (ICT) in Africa in various sectors of the economy including disaster management.

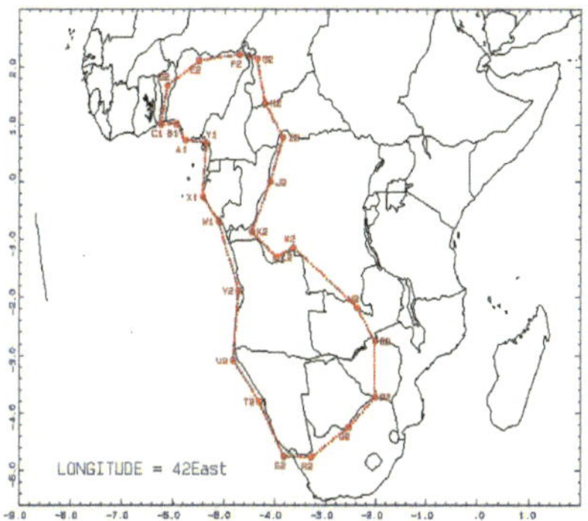

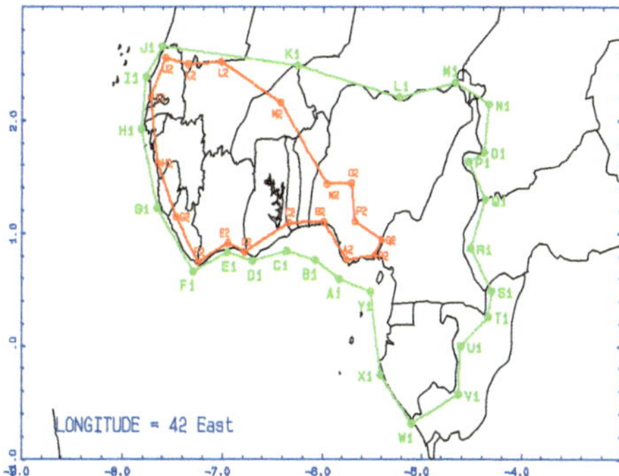

Fig 3.2 Ku-Band Ecowas-2 Antenna Beam Nigcomsat Project Handbook, 2006)

Fig 3.3 Ku-Band Ecowas-1 Antenna Beam (Courtesy

Although, Nigcomsat-1 experienced a power subsystem anomaly and was de-orbited within the period of 18 months for which it operated. Pilot projects on tele-medicine and tele-education were also demonstrated using bandwidth from NigcomSat-1. The technical partner in NigcomSat-1 project, China Great Wall Industrial Corporation, is working hard to produce another satellite (NigcomSat-1R) as a replacement within the next two years. Meanwhile, Nigcomsat Limited still offers services on third party satellites which will eventually be transferred to Nigcomsat-1R when completed (Ahmed Rufai, 2009). In future, the launch of NigComsat-2 and NigComsat-3 is intended to provide in orbit sparing for Nigcomsat-1R and cater for public services uses in areas such as disaster management.

Part from Satellite communications, the use of GNSS is beginning to gain relevance in Africa. This space technology has an enormous potential to aid disaster control in Africa. A good example of GNSS application in Africa is a regional development programme known as the Africa Reference Frame (AFREF).

3.4.6 Africa Reference Frame

The AFREF is a geodetic project designed to unify a co-ordinate reference frame for Africa and is based on network of permanent GNSS stations.

 The African Geodetic Reference Frame (AFREF) was conceived as a unified geodetic reference frame for Africa. It is to serve as a basis for the national and regional three-dimensional reference networks which would be fully consistent with the International Terrestrial Reference Frame (ITRF). ITRF is the global reference frame system for the earth as adopted by the International Association of Geodesy (IAG). When fully implemented, it will consist of a network of continuous permanent GPS stations such that a user anywhere in Africa would have free access to GPS data and products and would be at most 1000 km from such stations. Its full implementation will allow GNSS data from a network of permanent GNSS base stations are to be used for geophysical applications, weather forecasting, climate monitoring space weather monitoring and disaster mitigation (AFREE, 2007).

CASE STUDY OF NIGERIA SPACE APPLICATION IN DISASTER MANAGEMENT

4.1 Introduction

Over the years, Nigeria has been ravaged significantly by various types of natural disasters such as floods, drought, landslides, coastal erosion, sand-storms, dust-storms and locust infestations. Beyond these events, threats posed by frequent oil spills and irreparable damage to coastal biospheres, increasing levels of industrial pollution, waste and unprecedented climatic changes, and its negative consequences make Nigerians to be increasingly at risk to a wide number of new and emerging hazards. In this chapter, the main types of disasters in Nigeria and the modest attempt by NASRDA to mitigate these disasters via satellites would be reviewed. Particularly the chapter would further focus on monitoring of gully erosion in south east Nigeria using remote sensing technique, flood disaster risk assessment of Katsina Ala river basin and the control of desert encroachment in north eastern Nigeria.

4.2 History of Disaster Management in Nigeria

Organized Disaster Management in Nigeria dates back to 1906 when the Fire Brigade was established with functions that went beyond fire fighting to the saving of lives and property and provision of humanitarian services during emergencies. This situation continued until 1972 when Nigeria experienced a devastating drought. The drought had negative socio-economic consequences and cost the nation the loss of many lives and properties. This event amongst others led to the establishment of the National Emergency Relief Agency (NERA) in 1976 and was charged with the task of collecting and distributing relief materials to disaster victims. Considering the limited scope of NERA and based on the United Nations (UN), International Decade for Natural Disaster Reduction (IDNDR) vision, in 1990 Government set up an Inter-Ministerial body to address natural disaster reduction strategies. Thereafter, in 1993, the Government decided to expand the scope of managing disasters to include all areas of disasters. Furthermore in 1999, the role of NERA was expanded and the name of the Agency was changed to National Emergency Management Agency (NEMA) *(NDMF, 2009)*.

The general increase in the population of Nigeria in the last two decades has placed more people at risk in the event of extreme weather conditions. Also, the significant increase in human settlement particularly on floodplains over the past thirty years has increased the risk of flooding. If these trends continue, the costs associated with natural disasters will continue to be on the increase.

4.3 Disaster Management in Nigeria via Satellite

In the past, lack of understanding of the relevance of space acquired data, inadequate efforts to address the information needs of various economic sectors including the analysis and application of such information and the scarcity of skilled and educated man-power in space science and technology, were among the factors that confronted sustainable disaster management efforts in Nigeria through space. In recent times however, satellites remote sensing and other space systems are beginning to gain relevance for disaster mitigation in Nigeria. Aerial maps from satellite images have been produced for risk assessments. For instance, some areas that are prone to flooding have been mapped out by NEMA, thus allowing for better targeting of endangered people for purposes of early warning and response. For instance the map of Nigeria below shows areas prone to flood and Erosion was recently produced by NEMA GIS laboratory in Abuja

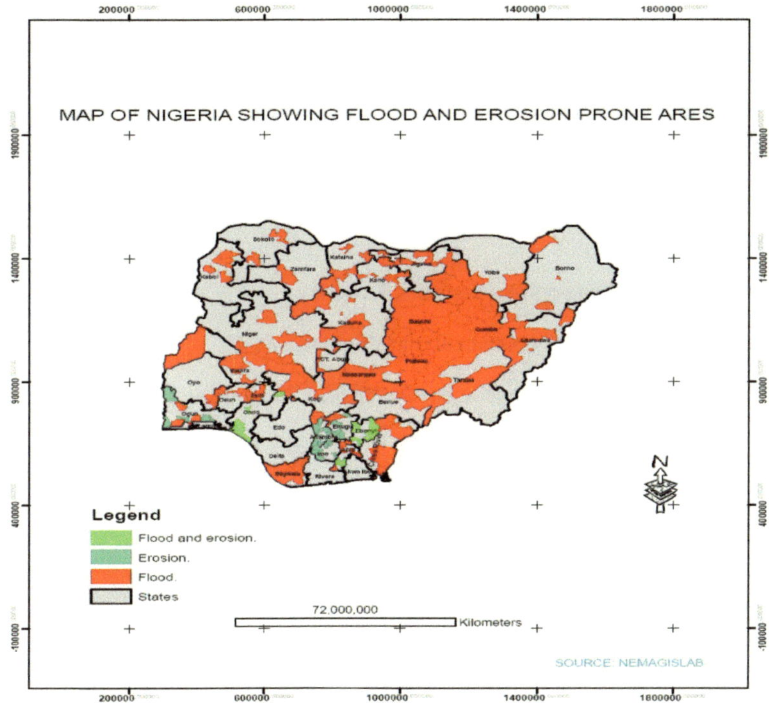

Fig 4.2: Map of Nigeria showing Flood and Erosion Prone Areas *Courtesy: NEMA GIS LAB*

According to immediate past NEMA's Director General Mohammed Audu-Bida, "Nigeria's NEMA was nominated by the International Charter on Space and Major Disasters (ICSMD) to co-ordinate disaster management in West Africa". The Director General, made this assertion at a mock exercise for the activation of the Charter. He described the charter as a multi-satellite operational set-up aimed at providing space-borne data on disasters that might have the potential of causing significant loss of life or property.(M Audu-Bida, 2010) He further added that the agency would give a wider application of space based technology for responsive disaster management.

4.4 Prevalent Hazards in Nigeria

The common hazards in Nigeria include:

- Frequent oil spills; pipe line vandalization in Niger Delta.

- Severe floods, especially in Jigawa, Kano, Sokoto, Kebbi, Zamfara Gombe and some Southern States.

- Threat of Desertification Droughts and general land use degradation the North Eastern axis of Yobe Borno and Adamawa States.

- Gully erosion traditionally in South Eastern states and becoming pronounced in Auchi and Bida.

- Wind storms in the northern parts of the country.

- Fire disasters especially market infernos Sokoto, Jos etc

- Cases of collapsed buildings in cosmopolitan cities such as Lagos, Abuja and Port Harcourt.

- Ethno-religious conflicts in Jos.

- Threat to oil and gas explorations by militia Niger Delta

4.5 Challenges of Disaster control

- Size and diversity of Nigeria.

- Lack of proper record.

- Ineffective coordination of disaster management efforts.

- Low Level of preparedness at government tiers.

- Inadequate awareness and advocacy on disaster preparedness.

- No policy on disaster data sharing and management.

4.6 Prospects of Disaster Control

- Numerous and willing stakeholders. NASRDA, Armed Forces, NMA, NGOs

- Nigeria acceded to UN Space Charter and collaborates with global disaster management initiatives e.g UN-SPIDER.

- Use of space technology.

- Disaster policy and structures through various organizations.

- NEMA placement directly under the Vice President of the Federal Republic of Nigeria.

4.7 Recent Effort

- Grass-root awareness creation.

- Simulation and training for stakeholders.

- Development of frame work for research.

- Enhancing forecasting coordination between NEMA and NASRDA.

- International collaboration through a regional body such as NEPAD.

- Involvement of Journalist and school clubs in disaster control awareness.

This report will now review the main types of disasters in Nigeria and the modest attempt by NASRDA to mitigate these natural hazards.

4.8 Monitoring of Gully Erosion in South East Nigeria

Gully erosion is an accelerated process whereby soil is displaced at a faster rate than it being replaced. Flood water is a major cause of erosion which is prevalent in South-Eastern Nigeria. The menace of gully erosion hampers the socio-economic life of the people living in the affected areas. In some cases human lives are lost to spreading gullies. Thus the knowledge of distribution of soil erosion in an area is important in planning land use and conservation measures.

The use of data from NigeriaSat-1 has created the opportunity to use remote sensing techniques to map and monitor the spread of gullies and also demonstrate the potentials of image data from NigeriaSat-1 in mapping and monitoring of gully erosion spread in South-Eastern Nigeria, with particular reference to Anambra State. The project generated some environmentally profitable data to help in modifying policies and decision-making processes relevant to land degradation. Thus the development of a methodology to access soil erosion is important in identifying affected and erosion prone areas. The project which focused on parts of South East Nigeria was implemented in collaboration with Nnamdi Azikwe University, Awka.

28

NigeriaSat-1, Landsat Enhanced Thematic Mapper plus (ETM+), SPOT 5, Quick bird images and Shuttle Radar Topography Mission (SRTM) data were used for the project. Thereafter, Landuse maps were generated for each state in the region and the gully erosion was classified according to its severity and the classification maps generated. Factors such as undulating topography, poor soil, poor agricultural practices, unplanned settlement pattern and urban development and sand mining were identified to aid gully development. Some socio-economic implications of the impact were identified. The study further recommended the development of Geo-information Based Early Warning System (GEOBEWS) which will be based on periodic acquisition of very important datasets that could be used to predict the possible development and expansion of gullies.

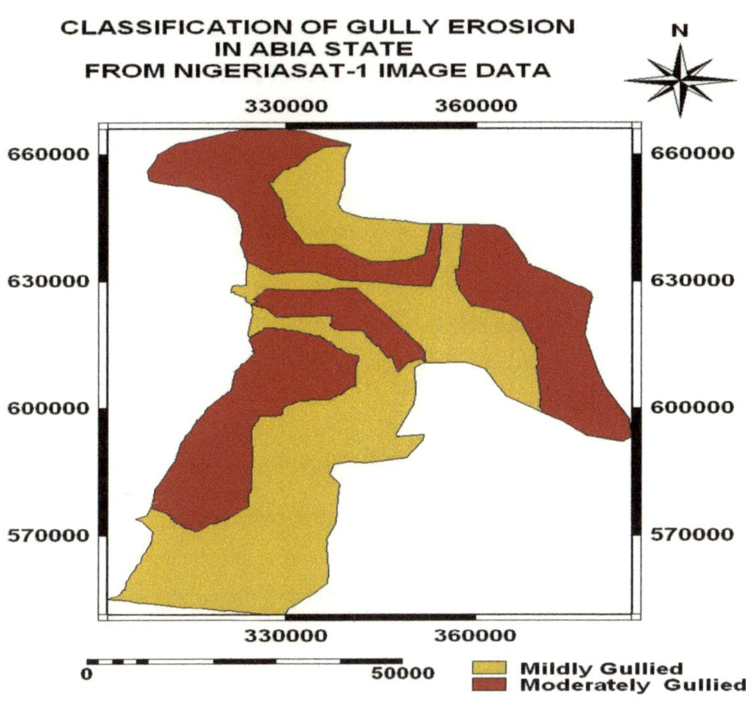

Fig 4.4 Classification of Gully Erosion in Abia State (Courtesy: NASRDA GIS Lab)

4.9 Flood Disaster Risk Assessment of Katsina-Ala River Basin

River Katsina-ala in Nigeria takes its source from the foot hills of Lake Nyos, located in the North-West area of Cameroon. This lake released poisonous gases (carbon dioxide) into the atmosphere in 1986, killing about 1800 people and about 3500 livestock. Recent study shows that erosion of the natural barrage that keeps the lake's water could lead to its collapse and the release of the water could lead to flooding that may affect the environment, settlement, agricultural land and humans along the flood plains of the drainage valleys especially the Katsina-ala river basin in Nigeria.

NASRDA carried out a space mapping on the Katsina-ala river basin using Nigeriasat-1. Thereafter, a hydrologic Modelling of the river catchment was done after carrying out a verification exercise during the dry season. Consequently, it was ensured that all settlements and resources at risk were adequately inventoried and cautioned.

4.10 Flood Vulnerability Assessment of Shiroro Dam

NigeriaSat-1 data has been used in a research on Shiroro Dam on monitoring of its water levels to reduce the potentials for flood disaster. The project was implemented in collaboration with the Federal University of Technology, Minna and has been completed since 2005.

4.11 Desert Encroachment in North Eastern Nigeria

Desert encroachment is a very serious environmental problem in the Northern Eastern part of Nigeria. Sand dunes are threatening to practically eat up the entire area, sparing nothing on their path. It is estimated that about 350 – 400 hectares of land is lost to desertification annually, with villages and access roads buried under sand dunes in some cases, Already villages like Tolutoluwa, which are close to the Niger/Nigerian border, are practically sitting on sand dunes. With virtually no farmlands in the vicinity and with livestock having to trek long distances before they can get places to graze. Thus, the inhabitants of places like Tolutoluwa are threatened by severe drought.

Fig 4.9: Sand dune in Tulotulowa

Consequently, NASRDA carried out a research to curb the southward expansion of the Sahara desert. The research was carried out in conjunction with the National Center for Remote Sensing (NCRS), Jos and the department of Surveying & Geoinformatics, Federal University of Technology, Yola (FUTY). The north eastern Nigeria was the region chosen for the research. In order to have a realistic and accurate assessment of the situation on the ground the North East Arid Zone Development Program (NEAZDP), an agency with sufficient knowledge of the local terrain, was also incorporated.

The area of interest (AOI) used in the project extends from latitude 11° 00'N to 13° 45'N and from longitude 10° 00' E to 13° 55'E. This area commonly known as the Sahel savannah covers about four states namely, Borno, Yobe, Bauchi and a little bit of Jigawa.

From the result of the research, NASRDA set up a database using data from NigeriaSat-1 and other satellites in combination with other climatic factors to model and determine the rate and parameters for desertification. The scheme created a Remote Sensing (RS) and Geographic Information System (GIS) Predictive Models for Early Warning against Desertification in Nigeria.

The project which was completed 2007 had the following objectives:

- To classify and evaluate the extent of desertification in the North Eastern Nigeria.

- To evaluate the vulnerability of the areas bordering these study area zones.

- To determine the rate of desertification over some years.

- To create a database on desertification.

- To produce of environmental awareness maps of North Eastern Nigeria.

- To produce a predictive model for desertification early warning.

4.11.1 Climatic Effect of Desertification

The elements of climate played very important roles in dictating the nature of the ground cover in the study area. Temperatures are usually very high reaching as high as 40.5°C in February/March. Lowest temperatures (as low as 26.8°C) are usually observed in December/January. s

5. CONCLUSION AND RECOMMENDATIONS

Disaster Management involves the coordination and integration of all activities necessary to build, sustain and improve the capability to prepare for, protect against, respond to, recover from, or mitigate against threatened or actual natural or man-made disasters. Due to the vantage position of satellites in orbit, space technology has proved to be an essential tool which Africa must fully utilize for disaster management.

Auspiciously, Africa has been receiving a lot of global attention in various aspects of disaster control, by reputable international organizations taking advantage of space systems in managing disasters in the continent. This global initiatives portends some opportunities and prospects for the development of space technology in Africa, particularly as it relate to disaster management. Therefore, African national governments should to support these international space initiatives by setting up platforms and activities aimed at creating access to space-based information for disaster management. Effort could also be made to bridge the gap between global disaster management institutions and existing African space centres so as to facilitate capacity-building and institutional strengthening in space disaster management. Furthermore, accessible linkages should be established to existing facilities and archives so as to improve the environmental sustainability of the continent.

The effort of NASRDA to use Nigeriasat-1 for disaster management in Nigeria is a laudable attempt which other African countries must emulate. The products from Nigeriasat-1data has been of benefit to programmes and policies of the Nigerian government in the area of disaster prevention. However, more attention could still be given to products that are essential for the protection of forest resources and support of livelihood of rural communities in Nigeria.

Finally, African countries must consider synergy of effort in space technology application development by pooling together resources at national and regional levels. In addition, there is a need to develop and enhance the present infrastructure especially in the areas of remote sensing and GIS including, the organizational and computational data processing, particularly as relates to anticipated disaster mitigation needs in Africa.

List of References

AARSE, 2010 African Association of Remote Sensing of the Environment, http://www.itc.nl/aarse/aboutus.html. (Accessed on 24 June 2010)

ADEPOJU, K. 2010, Resident Faculty, ARCSST-E, Personal Communication July 2010.

AMBROSE, S, HABIB, S, and MCKELLIP R, 2005. Extending NASA Research Capabilities for Disaster Management. Earth Observation Magazine August 2005
[online]:from:http://www.eomonline.com/EOM_Aug05/article.php?Article=feature01
(Accessed 11 August 2010)

AUDU BIDA M, Arique Avenir Jornal May 2010, www.AfriqueAvenir.org accessed o 13 July 2010.

BERNARDI M et al, 2000, World Meteorological Organization Expert Group Meeting on Software for Agroclimatic Data Management Washington, DC, USA, 16-20 October 2000

BHANUMURTY V & SHRIVASTAVA NK, ISRO 2010, Paper on Collaboration with International disaster Charter presented at Report Session: Setinel Asia, Bankok.

BOROFFICE R.A & AKINYEDE JO 2005, Space Technology and Development in Africa and the Nigerian Experience, 2005, p36.

CIS,2008, Common Information Space of the United Nations Organizations in Bosiann "UN in Bonn - Working Towards Sustainable Development Worldwide.p.24 http://www.unric.org/html/german/UN-in-Bonn.

COSPAS-SARSAT, International Satellite System for Search and Rescue [online]. Montreal. Available from: http://www.cospas-sarsat.org/MainPages/indexEnglish.htm
[Accessed 9 August 2010].

UIC 2009, Disaster Risk Management and Emergency Response - Reaching The End Users,GEO User Interface Committee 13th UIC Meeting • Washington, DC 15 – 16 November 2009.

FAO. 1995. "Crop production System Zones of the IGAD sub-region". FAO Agrometeorology, Working Papers Series No. 10, by Van Velthuizen, H., L. Verelst and P. Santacroce. FAO,Rome.

GMFS Operation Report 2010, Earth Watch GMFS Service Element. http: www.gmfs.info/uk/publications. Accessed on 3 August 2010.

LEWIS, S. 2009, Remote Sensing for Natural Disasters: Facts and figures, Article Published on Science and Development Network, November 2009, http:fews .net (Accessed on 2 August 2010).

MOMOH J AND AKINYEDE J, 2008. African Regional Challenges and Position in Space-Based Disaster Management and Reduction. Article Published on African Skies, Oct 2008 Edition.

NDMF, 2009 National Disaster Management Framework. A publication of NEMA, 2009 p4

Nigeria Communication Satellite Project Handbook 2006. A publication of Nigcomsat Limited Abuja 2006.

Space Application Projects Summary 2007. A publication of National Space Research and Development Agency, Nigeria. 2007

United Nations Resolution 61/110
http://www.unoosa.org/pdf/gares/ARES_61_110E.pdf (Accessed on 19 June 2010)

WAHLSTROM, M. 2010, Improving Disaster Management and Crisis Response in Africa, an Article on Defence IQ, May 2010 edition.

YOUR KNOWLEDGE HAS VALUE

- We will publish your bachelor's and
  master's thesis, essays and papers

- Your own eBook and book -
  sold worldwide in all relevant shops

- Earn money with each sale

Upload your text at www.GRIN.com
and publish for free